WAWG
is pleased to sponsor the publication of this commemorative photo book

IMAGES OF THE LAND
Washington Wheat Country

Proceeds from the sale of this book will augment WAWG's Legislative Action Fund.

The Washington Association of Wheat Growers (WAWG) is sponsoring this book because it highlights the beauty of the wheat land we care for, and because part of the proceeds from book sales will go directly into WAWG's Legislative Action Fund (LAF). LAF funds are specifically targeted at legislative activity that directly affects Washington wheat growers. In many cases, for example, one or more people need to travel to Washington, D.C., or Olympia to present our side of one of many stories—supporting wheat growers as they carry out their ongoing business.

We are appreciative that you have chosen to help us in our important work, while taking a refreshing photographic journey with *Wheat Life* Editor David Andersen.

ENERGY SAVER
BLMR 1015
GRAIN
TRAIN
AT WORK FOR
WASHINGTON STATE
ENERGY SAVER
BLMR
1017
GRAIN
TRAIN
AT WORK FOR
WASHINGTON STATE
ENERGY SAVER
BLMR 1016
GRAIN
TRAIN
AT WORK FOR
WASHINGTON STATE

Washington Association of Wheat Growers Celebrates Its Fiftieth Year!

I am blessed to wake each morning in one of the most beautiful wheat-growing regions of the world. That region is the Pacific Northwest and, specifically, I'm referring to the eastern portion of Washington state. Great beauty is found within the diversity of eastern Washington. In this area east of the Cascade Mountain Range, wheat is grown in locations that receive as little as six inches of precipitation to areas that receive more than thirty inches annually. The terrain and soil types vary as much as the precipitation. Farming practices change from one location to another, and the separation of these locations in some cases can be as little as a few miles.

Fifty years ago, a group of innovative farmers joined together to form the Washington Association of Wheat Growers. The vision of this organization's charter members has led to huge advancements in the wheat industry and has provided wheat producers with a strong, unified voice. This unified voice was important then and will be vital to our wheat producers' future.

In an effort to honor those who have served our industry, we decided to publish a book of photos that showcase the beauty of our area. The images presented in this book will help people unfamiliar with our land understand why farmers are passionate about their livelihoods. Farming is so much more than just a job!

—Jeff Emtman, President, 2005
Washington Association of Wheat Growers

STATE OF WASHINGTON
DEPARTMENT OF AGRICULTURE
PO Box 42560 • Olympia, Washington 98504-2560 • http://agr.wa.gov • (360) 902-1800

Coming from an agricultural family with roots in eastern Washington, I am impressed with the variety of agricultural commodities produced by Washington farmers and ranchers—we represent one of the most diverse agricultural and trading states in the United States.

Washington's wheat growers can be proud of the contributions that they have brought to this state for more than 150 years. Wheat is the third-highest-ranking agricultural commodity with a production value of $521 million. And Washington ranks number five among U.S. wheat-producing states.

Washington is one of the nation's largest wheat-exporting states, and I'm proud of the services that the Washington State Department of Agriculture (WSDA) provides to help make wheat exports successful. We sample, inspect, and weigh grains to ensure a consistently high-quality product is shipped to your export markets.

I am never more optimistic about the future of Washington agriculture than when I return from a trade mission overseas. We work to promote our state's products and it's paid off in gains for our producers. WSDA continues to be an advocate to help wheat growers meet their challenges.

The photos in this book will make it obvious how the wheat industry, and the people who labor in it, have made great contributions to our landscape and our economic diversity. I am proud of the wheat farmers and their families who nurture the land and support our communities across the state.

Sincerely,

Valoria H. Loveland
Director

IMAGES OF THE LAND
Washington Wheat Country

by
David Andersen

THE
DONNING COMPANY
PUBLISHERS

All photos taken by David A. Andersen, Editor, ***Wheat Life*** magazine, for use in ***Wheat Life*** magazine, with permission given for reprinting in the book titled ***Images of the Land: Washington Wheat Country***, 2005

The Donning Company Publishers
184 Business Park Drive, Suite 206
Virginia Beach, VA 23462

Steve Mull, General Manager
Barbara B. Buchanan, Office Manager
Richard A. Horwege, Senior Editor
Bobbi LaChance, Graphic Designer
Stephanie Bass, Imaging Artist
Mary Ellen Wheeler, Proofreader
Cindy Smith, Project Research Coordinator
Scott Rule, Director of Marketing
Stephanie Linneman, Marketing Coordinator

Dwight Tompkins, Project Director

Library of Congress Cataloging-in-Publication Data
Andersen, David.
Images of the land : Washington wheat country /
by David Andersen.
p. cm.
ISBN-13: 978-1-57864-330-1 (hard cover : alk. paper)
ISBN-10: 1-57864-330-9 (hard cover : alk. paper)
1. Wheat—Washington (State)—Pictorial works. 2. Agriculture—Washington (State)—Pictorial works. 3. Washington Association of Wheat Growers. I. Title.
SB191.W5A568 2005
633.1'109797022—dc22

2005028228

Printed in the USA at Walsworth Publishing Company

A Hardy People, Dedicated to the Land

This book is dedicated to Washington state wheat producers, their families, and the generations before them who built an industry that helps feed people around the world. The people of this industry also provide America with the strong cultural fiber, or sinew, that has helped bind America together since its beginnings. I have found these people to be sure yet cautious, frugal yet generous, and devoted to the land they husband. As a WAWG past president has said, "These people are the salt of the earth," and their care of the land has deeply impressed me. I hope these images portray the beauty of lands cared for by such people.

While carrying out multiple roles as editor of WAWG's award-winning monthly magazine ***Wheat Life***, I have usually traveled with camera equipment and film, stopping to compose photos en route to various meetings and conferences. I did not have time to find and make notations about photo locations, which resulted in achieving my goal—to show the breadth of beauty of the entire state wheat-growing region. If the photos speak about the beauty of Washington wheat country, they have been included in ***Wheat Life***, and are included herein. Showcasing that beauty as a whole has been and remains my goal.

This book is intended not just for state wheat producers, landlords, and supporting businesses, even though I hope these people take it to heart as much as I have taken the photos with heart. We know the beauty of this land—you who grew up on it and newcomers like me. It is also intended to be shared with others who live outside of our agricultural world in and out of Washington state, to help them understand that supporting their personal lifestyles—their SUVs or boats or summer homes—is a bushel of wheat. We're all in this great country together. The farmers of America are holding up their end in the equation, thus keeping America strong. We should thank them and honor them. This book is one such attempt.

—Thank you, David A. Andersen, Author, 2005
Editor, ***Wheat Life*** magazine

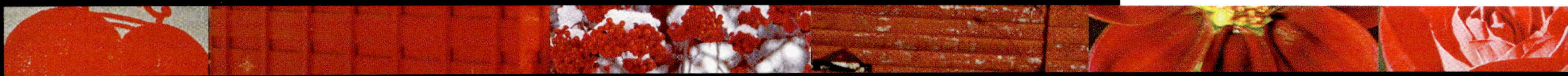

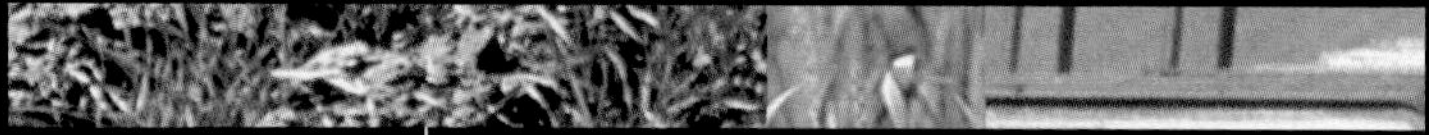

Washington Wheat Country

Washington wheat country comprises nearly half of the state of Washington, in thirteen contiguous counties that fill the bulk of the land in the east and southeast. People are familiar with the wet green side of Washington but, once over the Cascade Mountains and heading east and south, a visitor discovers a unique beauty—drier, and in most areas cultivated with caring farm practices. It is a land of plenty, of splendor in the full-bore sun. In that sun, some of the best wheat in the world is grown.

Because the rain is blocked and absorbed on the west side of the Cascades, the east side thrives on rains of eight to twelve inches, to as much as twenty-two to twenty-four. Wheat farms that depend on sky-fed moisture are called dryland farms, and often employ fallowing, or leaving some of the land to absorb moisture only, following which they are cultivated, seeded, and harvested again.

According to our good friends at the Washington Wheat Commission, "Many wheat farmers have switched to conservation tillage practices to further augment soil moisture and reduce topsoil loss due to erosion. Annual cropping (as opposed to summer fallow rotation) means the ground is planted yearly. It is mainly used in regions with higher rainfall and on irrigated farms. A kaleidoscope of terrain, soil types, and weather patterns throughout the state means wheat yields vary widely by region. One dryland farm can have fields that yield as little as twenty-five bushels per acre to as much as one hundred." (From the commission brochure "Washington Wheat Facts.")

As you read ***Wheat Life*** magazine and gain information from other sources you will see that beneath the diversity and awesome beauty are layers upon layers of complex farming operations and issues, all of which are meant to support farm families, continue to add to the state's economic strength, and help America maintain its national security. Most prominently, you will see that Washington wheat farming is inextricably linked to the nation through the federal government.

But for now, enjoy a virtual, visual ride through Washington wheat country.

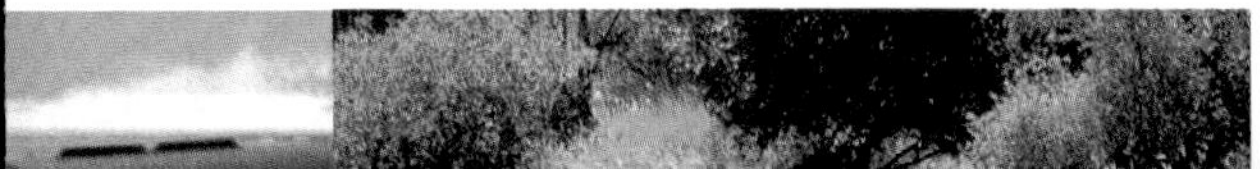

JOHN DEERE

203
WELCOME

400
AGCO

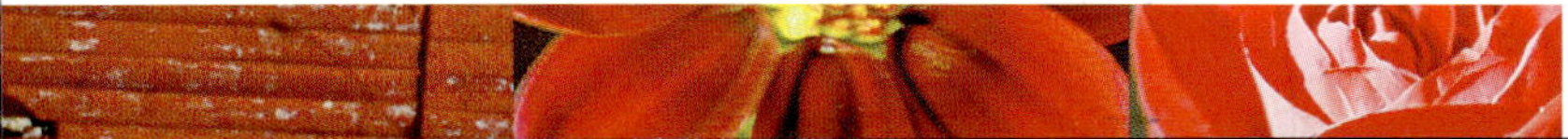

U.S. POST OFFICE-STEPTOE WA. 99174
COOPERS' CORNER
USED
ANTI
YES WE'RE OPEN

GLEANER

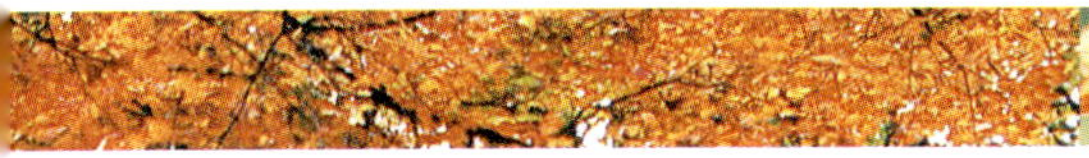

BLUEGRASS
50 LBS.NET

W
E

RAIL ROAD
CROSSING
R R

FORD

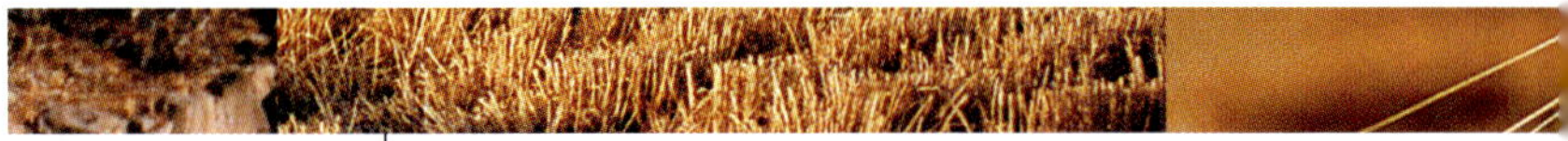

MT

FARMER HALL

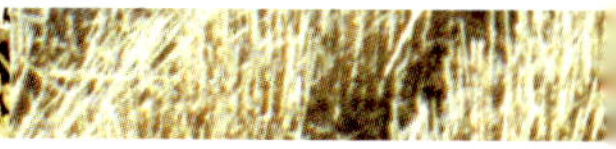

3